Paulo Venâncio de Sousa
Adna Cristina Barbosa de Sousa

ARTISANAL PRODUCTION OF WHITE SHIMEJI

Paulo Venâncio de Sousa
Adna Cristina Barbosa de Sousa

ARTISANAL PRODUCTION OF WHITE SHIMEJI

Axenic cultivation of Pleurotus ostreatus var. Florida on co-products generated from agricultural production

ScienciaScripts

Imprint

Cover image: www.ingimage.com

This book is a translation from the original published under ISBN 978-3-639-69463-5.

Publisher:
Sciencia Scripts
is a trademark of
Dodo Books Indian Ocean Ltd. and OmniScriptum S.R.L publishing group

120 High Road, East Finchley, London, N2 9ED, United Kingdom
Str. Armeneasca 28/1, office 1, Chisinau MD-2012, Republic of Moldova, Europe
Managing Directors: Ieva Konstantinova, Victoria Ursu
info@omniscriptum.com

Printed at: see last page
ISBN: 978-620-8-64055-2

ACKNOWLEDGEMENTS

To the FEDERAL UNIVERSITY OF PARAÍBA *for the fundamental support, encouragement and technical-scientific assistance that was essential for carrying out this research.*

To LABORATORY OF GENETICS MOLECULAR AND BIOTECHNOLOGY

VEGETAL (L G M BIOTEC) *for generously providing the physical space to carry out this study.*

To the CENTRE OF BIOTECHNOLOGY *in its entirety, including all its employees and students, for providing the essential infrastructure to carry out this work, as well as being an environment that marked an enriching and learning phase in my life.*

I would like to express my sincere and profound thanks to ***PROF. DR. ADNA CRISTINA BARBOSA DE SOUSA*** *for the opportunity offered to carry out the Voluntary Institutional Programme for Scientific Initiation and Course Conclusion Work, required at the end of undergraduate courses as an advisor and teacher, but also for her patience, kindness and welcome throughout the process.*

I thank God, the *source of wisdom and strength, for guiding my steps and giving me the grace and perseverance to complete this stage of my life.*

SUMMARY

Pleurotus ostreatus var. florida's ability to colonise and degrade a variety of lignocellulosic residues has offered several benefits, including waste reduction and the production of nutritious food in a sustainable way. Several residues are generated from the bean and algaroba crops, such as pods after threshing the beans and leaves after removing the algaroba pods, most of which are disposed of improperly. Therefore, with a view to reusing this waste, the aim of this study was to evaluate the development of P. ostreatus var. florida in a new alternative compost from bean pods and algaroba leaves. Biological and production parameters and carbon (C), nitrogen (N) and C;N ratio contents were assessed in six different formulations with proportions ranging from 5% to 50% algaroba leaves and from 50% to 95% pods. Based on the results, the total colonisation time for mycelial development in the substrate ranged from 15 to 23 days. The fastest mycelial development was observed in TC (11.50cm), followed by treatments T1 (9.10cm) to T6 (8.80cm), which did not differ statistically. With regard to biological parameters, the total cycle from vegetative and reproductive development varied from 21 to 23 between the 6 treatments. Stipe size ranged from 1.67 to 2.00 cm. In relation to the area of the pileus, T2 (12.74cm$^{2)}$, T3 (14.31cm^{2}), T4 (10.74cm$^{(2)}$)) and T5 (10.57cm$^{(2)}$)) developed more abundantly. With regard to the production parameters found in the different treatments, it was observed that basidiom moisture varied from 89.00 to 91.00%, yield from 12.50 to 18.00%. Its biological efficiency ranged from 39.06 to 56.25%. Organic matter loss ranged from 28.75 to 48.74% and compost consumption from 51.25 to 71.25%. The C:N ratio between the 6 formulations was approximately 30:1 to 36:1. Based on the results obtained, the formulations with the highest yield and biological efficiency were T1, T2, T3 and T4. They were therefore promising for mushroom cultivation, as they were able to sustain continuous vegetative development and basidiom formation.

Keywords: *Mushrooms, lignocellulose, food, reuse, yield.*

SUMMARY

1 INTRODUCTION

Pleurotus ostreatus var. florida is an edible mushroom belonging to the Pleurotaceae family. This mushroom is widely recognised for its pleasant taste, high nutritional value and medicinal properties. Mushrooms have been known for more than three thousand years by Asian peoples and have been used both in traditional medicine and in food as an excellent source of protein (URBEN et al., 2017).

Since the 1970s, their consumption has increased, mainly due studies suggesting that their nutritional and medicinal properties provide health benefits (URBEN et al., 2017). The mushroom market is already worth 35 billion dollars a year worldwide (GOMES, 2018). This market is expected to grow in the coming years (CAPRA; TONIN, 2019).

In Brazil (1996), each Brazilian consumed an average of 30 grams of mushrooms a year (SANTOS, 2018). Today, the average annual consumption has risen considerably to 160 grams (GOMES, 2018). Until the early 2000s, due to low production and high cost, mushrooms were not part of the culture and diet of the Brazilian people (URBEN, 2017).

The cultivation of P. ostreatus var. florida using vegetable waste, such as bean pods (Vigna unguiculata L.) and algaroba leaves (Prosopis juliflora), presents a promising opportunity to increase the production of alternative foods in an economically and environmentally responsible way. These agricultural residues contain nutrients that can be converted into high-quality mushroom biomass. As well as providing economic and environmental benefits, this practice can contribute to a more diverse and nutritious diet and could be a promising strategy for producing alternative and sustainable foods.

Evaluating the growth and vigour of P. ostreatus var. florida on lignocellulosic substrates such as bean pods and algaroba leaves is crucial to understanding the potential of this combination as a growing medium. Parameters such as substrate colonisation rate, fruiting time, mushroom yield, and nutritional quality of the fruiting bodies are important indicators for determining the effectiveness and viability of this approach.

Implementing this practice can have a positive impact on the environment, promoting agricultural sustainability and reducing pressure on natural resources. Based on this, the aim of this study was to evaluate the cultivation and development of P. ostreatus

var. florida in new alternative compost from bean and algaroba crops.

2 OBJECTIVES

2.1 GENERAL OBJECTIVE

To evaluate the cultivation of P. ostreatus var. florida in a new alternative compost from bean and algaroba crops.

2.2SPECIFIC OBJECTIVES

✓Evaluate the potential for using bean pods and algaroba leaves to formulate a substrate for mushroom production;

✓ characterise the biological parameters: substrate colonisation time, primordia emission period, basidiom formation time, basidiom diameter, basidiom height, etc. pileus and pileus width;

✓Characterise the production parameters: yield, biological efficiency, productivity, compost consumption and loss of organic matter of the different formulations;

✓Carry out the chemical characterisation of the formulations in terms of carbon (C), nitrogen (N) and C:N ratio;

✓To classify the best formulation for vegetative development and production of fungal biomass.

3 THEORETICAL BACKGROUND

3.1 Mushroom production in the world and in Brazil

Since their origins, humans have mainly lived in a hunter-gatherer society, depending on the biodiversity of their surroundings for their sustenance. The history of mushrooms dates back to 300 BC, when Theophrastus first mentioned them as "truffles". The Romans considered mushrooms to be "food of the gods", while Greek warriors used to consume them during battles to gain strength (PANDITA, 2023). Today, mushroom cultivation is carried out all over the world, adopting the most diverse technologies possible, from the simplest to the most technologically advanced. It should be noted that the degree of economic investment and local social conditions in each country are factors that affect the technological level of cultivation. Despite this, the entire production process starts with the selection of the raw materials that will be used to produce the compost/substrate. Pleurotus spp. are examples of primary decomposers. These mushrooms usually grow rapidly, from primary mycelium that quickly attaches to and decomposes plant tissue. Each species has developed specific sets of enzymes to break down lignin-cellulose, the structural components of most plant cells, so that once the enzymes of one mushroom species have broken down lignin-cellulose to their full potential, other saprophytes will utilise their own repertoire (ZIED, 2021). The net result is exponential growth in global mushroom production. Due to the almost unimaginable growth in the production of shiitake, Pleurotus spp. (shimeji), wood ear mushroom and Flammulina, particularly in East Asian countries such as Japan, South Korea, Taiwan, China, etc.., FAOSTAT ((Food and Agriculture Organisation Corporate Statistical Database) is a robust database of crops, including mushrooms and truffles. According to FAOSTAT, the total world production of mushrooms in 2019 was 11,890.00 million. Countries such as China, Japan, India, Iran, the United States of America, Canada, the EU, the Russian Federation and Australia accounted for

98 per cent of the world's mushroom production. Most of the production came from Asia. Mushroom production in Asia in 2019, as reported by FAOSTAT, was 9,854.00 million tonnes. In Asia, China, Japan and India accounted for more than 97 per cent of production (SINGH, 2020).

At the beginning of this century, P. ostreatus was making the maximum contribution to total production in China. In the years since, shiitake production has soared and the increase has been more than fivefold in the last two decades. As a result, shiitake has moved from second place to first. Auricularia spp. moved from third to second and P. ostreatus from first to third position (SINGH, 2020). Although P. ostreatus was more popular at the beginning of this century, over the years other Pleurotus species are becoming more popular, the most produced being P. eryngii. In recent years, P. geesteranus has gained prominence, possibly due to its shape and flavour (SINGH, 2020). At the end of the 19th century, the USA was the main player in global button mushroom production. It was here the white strain was found and popularised, and technology for spawning grains and pasteurising compost was developed. Even today, the button mushroom contributes 98 per cent of total US mushroom production. After a continuous increase in the last century, button mushroom production has fluctuated in the present century at around 400 million kg and in the last five years, there has been a continuous decline, while there has been a continuous increase in production in Canada during these years. In the last five years (2016-2020), however, production has more than doubled, with Pleurotus spp. showing more growth than shiitake (SINGH, 2020).

The USA is a net mushroom importer (a country or region that imports more mushrooms than it exports. This means that in order to meet its domestic demand, the country buys more mushrooms from other countries than it sells abroad. This concept is related to the balance of trade: when mushroom imports exceed exports, the country is considered a net importer of this product). There has been a tenfold increase in mushroom consumption over the last five decades and current consumption is around 1.3kg/person/year. The USA, Canada,

Mexico and Brazil account for 95.3 per cent of mushrooms on the American continent. 92.9 per cent of mushroom production is accounted for by three countries: the USA (63.8 per cent), Canada (19.3 per cent) and Mexico (9.8 per cent). Mexico is considered the main mushroom producer in Latin America. Of the remaining Central and South American countries. Brazil, Chile and Colombia account for 5.4 per cent and the other countries account for the remaining 1.7 per cent (SINGH, 2020).

More than 92 per cent of mushroom production in Europe is carried out by European Union countries, although they only make up 19.1 per cent of Europe's area and 68.4 per cent of Europe's population. In the rest of Europe, Russia occupies the most space and accounts for 74.3 per cent of Europe's area and 19.1 per cent of its population (SINGH, 2020).

In 2019, Russia was the largest producer in the rest of Europe, followed by Belarus, from where production is mostly exported. Total production in the rest of Europe in 2019 was 110 million kg (SINGH, 2020).

Mushrooms are grown in most of Australia's states, close to population centres in peri-urban areas. The main production areas are near Adelaide in South Australia; Melbourne Metro in Victoria and Sydney Basin in New Wales. South. In 2019/20, total reported production was 68,823 tonnes, of which Victoria, New South Wales and South Australia accounted for more than 83% of production (SINGH, 2020).

Africa constitutes at least 25 per cent of the world's total mushroom biodiversity, but contributes only 0.4 per cent of total mushroom sales and new mushroom products on the global market. In South Africa, around 21,000 tonnes per year (2018) of mushrooms are produced with half of the mushroom growers in Gauteng, 28% in the Western Cape and 17% in KwaZulu-Natal. Some South African mushrooms are exported, mainly to neighbouring countries such as Namibia and Mauritius. Considering the growth patterns in the main mushroom-producing countries, it is estimated that global production could exceed 50 MT

(50,000 tonnes) by 2025 (SINGH, 2020).In Brazil, mushroom production began in the Alto Tietê region of São Paulo. Over the last few years, despite all the difficulties, the mushroom production chain in Brazil has been structured and mushroom cultivation is now widespread in various regions of the country. Currently, the main producers are in the states of São Paulo (Mogi das Cruzes, Pinhalzinho, Ibiúna, Sorocaba, Salto, Cabreúva, Juquitiba and Valinhos) and Paraná (Castro, Tijucas do Sul and Curitiba). In addition to these states, there is also cultivation in Minas Gerais, Rio de Janeiro, southern Bahia, Pernambuco, Brasília and Rio Grande do Sul (AMAZONAS, 2013).

The lack of precise estimates about the sector makes it difficult to quantify the exact number of active producers in Brazil and the ANPC (National Association of Mushroom Producers) estimates that there are more than 300 mushroom producers in the country, most of whom are micro and small family farmers. These producers are responsible for generating more than 3,000 direct jobs and have mushroom production as their main source of income (AMAZONAS, 2013).

Technological improvements have allowed Brazilian fungiculturists to gain in productivity in recent years, but for more than a decade a significant proportion of national producers (producers of champignon de paris) have suffered from the often unfair competition faced by national fungiculture in relation to imported products (provisionally preserved and pickled mushrooms), especially products originating in China (AMAZONAS, 2013).

In Brazil, mushroom production began in the Alto Tietê region of São Paulo. Over the last few years, despite all the difficulties, the mushroom production chain in Brazil has been structured and mushroom cultivation is widespread in various regions of the country. The main mushroom-producing states in Brazil include São Paulo: the main producer, especially of A. bisporus and P. ostreatus, with the emphasis on the regions where the mushrooms are grown. such as Mogi das Cruzes, which has a developed infrastructure for cultivation (URBEN, 2017). Minas Gerais: The second largest producer, with a focus on Pleurotus production. The state has seen an increase in cultivation on family farms (GOMES, 2018). Paraná: Has a growth in production, focusing mainly on the cultivation of Agaricus and some Pleurotus species (CAPRA, 2019). Bahia: A developing state in mushroom cultivation, with initiatives aimed at sustainability and the use of agricultural waste (SANTOS, 2020).

In 2020, the National Association of Mushroom Producers (ANPC) estimated that Brazil had approximately 1,500 mushroom producers. This figure represents significant growth compared to previous years, reflecting the increased demand for fresh mushrooms and awareness of the nutritional benefits of these fungi. The ANPC also pointed out that mushroom production in Brazil was diversifying, with a growth in the cultivation of different species, such as A. bisporus, P. ostreatus and Lentinula edodes, as well as the development of new cultivation techniques and the use agricultural waste as a substrate. This growth indicates continued expansion potential for the mushroom production chain in the country, especially with the growing demand for sustainable and healthy food (ANPC, 2020).

Due to a lack of tradition and ignorance mushrooms, their benefits and even how to prepare them, Brazil also has a low per capita consumption of around 160 grams, compared to some European countries such as France, Italy and Germany, whose per capita consumption is over 2kg, or Asian countries such as China and South Korea, which consume over 8kg of mushrooms per inhabitant (AMAZONAS, 2013). Within the Pleurotus genus, several species of mushroom are cultivated, including the Pleurotus ostreatus (white shimeji and black shimeji), P. djamor or P. ostreatoroseus, P. eryngii, P. pulmonarius and P. citrinopileatus. These mushrooms are grown on a range of organic substrates (cereal straw, sawdust, sugar cane bagasse, cottonseed co-products) in plastic bags or bottles. Mushrooms of the Pleurotus genus occupy the second position in world production, accounting for 25 per cent of it. In Brazil, this group also occupies the second position and it is estimated that its production represents more than 16 per cent of the total "in natura" mushrooms produced in the country

(AMAZONAS, 2013). In recent years there has been a slow change in the Brazilian fungiculture scene. Producers have been investing in techniques and professionalising mushroom cultivation in some regions. São Paulo, for example, has high-tech companies producing Pleurotus spp. or L. edodes (Mogi das Cruzes, Sorocaba, among others), and also in the south-west of Bahia (Vitória da Conquista and neighbouring areas). Although the national scenario is not very representative of the consumption of this product, there has been a significant increase in production and consumption in certain regions of Brazil. The South (Paraná, Santa Catarina and Rio Grande do Sul) is the country's main producing region, due to the large immigration of Asians to this region, thus implanting such consumption habits (MACHADO, 2020).

3.2 General aspects and nutritional characteristics of Pleurotus ostreatus var. florida

Fungi have a structural need for organic materials, either from non-living sources or from hosts. In the first case we refer to fungi as saprophytes, and in the second as parasites. They absorb nutrients, but commonly secrete enzymes that break down complex insoluble or less soluble nutrient materials that can be absorbed. This is sometimes called absorptive nutrition, or it is said that nutritionally, fungi are osmotrophs. Fungi have a cell wall, the main components of which are glucans and chitin (polysaccharides) (CHANG, 2020).

In structural terms, filamentous fungi have a tubular structure called a hypha. Through branching, and in some species through anastomosis or fusion of hyphae, a network of these filaments is formed, called a mycelium. Some species of filamentous fungi, such as the Basidiomycetes, are septate as shown in "Figure: 1" Various types of septa are found in the hyphae of fungi. A more complex septal structure is found in the Basidiomycetes. Here there is a tumescence around the pore region, and this is called a dolipore septum. Above these, on either side of the septum, are dome-shaped structures, the septal pore caps or parentheses, which are perforated membranes continuous with the endoplasmic reticulum. The origin of this type of septum and considerations about the mechanisms by which nuclei can migrate through such septa (CHANG, 2020).

Figure: 1 - Types of hyphae.

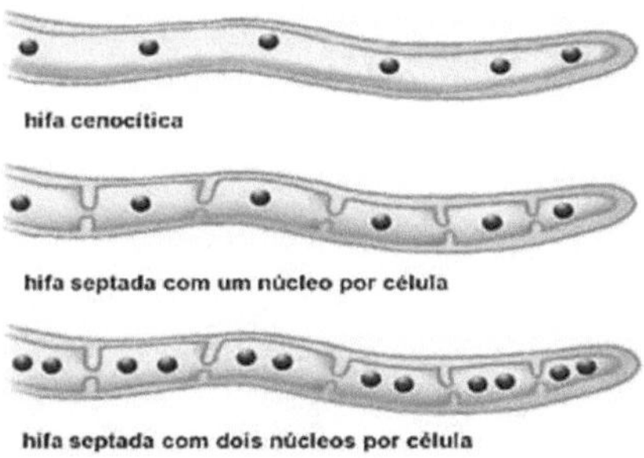

MELHOR DA BIO. Tipos de hifa. Melhor da Bio, 2016. Disponível em: https://melhordabio.blogspot.com/2016/12/tipos-de-hifa.html. Acesso em: 20 out. 2024.

As most cultivated edible fungi are Basidiomycetes, the mycelium derived from a single spore is called the primary mycelium, while the mycelium formed after the fusion of two compatible primary mycelia is called the secondary mycelium. The secondary mycelium is made up of hyphae, which have pincer-like connections in which the dikaryotic condition is maintained. The hyphae of the secondary mycelium maintain this dikaryotic condition by the process of clamping in the connection formation. When a migrating nucleus from a compatible lineage travels through a resident hypha and reaches a tip cell, the tip cell becomes dikaryotic, possessing two non-fused compatible nuclei. Near the position in the hypha where the two nuclei are located, a small protrusion of hypha appears and then grows away from the tip. This growth is the so-called hook cell. Before the hook cell recurves and merges with the main hypha, the two nuclei undergo a simultaneous (sometimes called conjugate) division, and one of the daughter nuclei moves into the hook cell. A daughter nucleus moves away from the tip to the region of the hypha close to where the hooked cell approached the main hypha. Compatible nuclei are located in the tip region of the hypha (CHANG, 2020).

The structure formed by the hooked cell is called a staple connection as shown in "figure: 2". This connection maintains the dikaryotic condition in the tip cell where growth occurs. Fusion of the nuclei does not normally occur until the fruiting body with its basidia are formed, and thus the dikaryotic condition (hyphae with clamp connections) is maintained in nature and can be perpetuated in culture for prolonged periods "figure: 3" (CHANG, 2020).

Figure 2: Clamp connection.

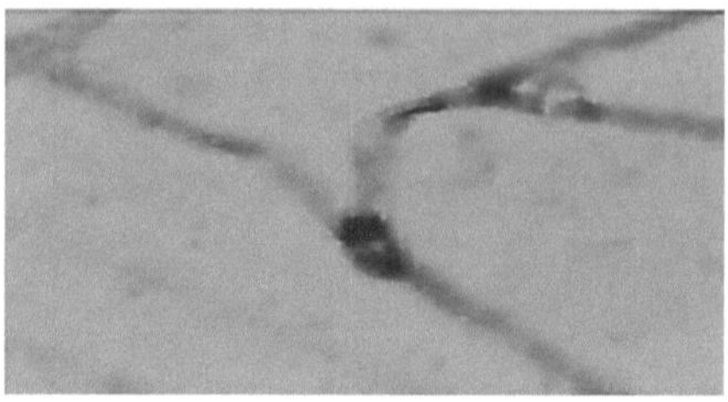

Source: CHANG, 2004).

Figure 3: Schematic representation of the formation of the clamp connection.

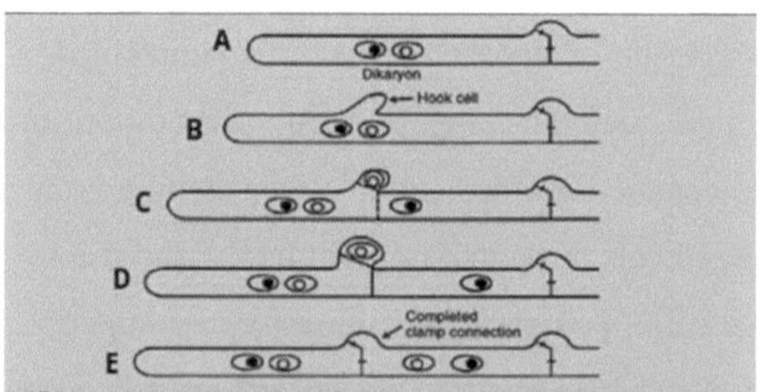

(A) Divisão nuclear conjugada ocorre. (B) Início da formação septal. (C) Formação septal concluída e fusão da célula em gancho com penúltima célula. (D) Passagem do núcleo da célula em gancho para a penúltima célula. (E) Conclusão da braçadeira conexão.

(CHANG, Shu-Ting, et al, 2004)

Two of the three basic stages of the sexual cycle occur in the fruiting body of Basidiomycetes. The fruiting body, which is the most notable stage in the life cycle of a basidiomycete, is a veil, a membrane that partially or completely covers the pileus (cap) of the mushroom when it is young. As the mushroom grows, the veil can break off, leaving fragments at the base of the stipe (such as the volva) or forming rings around the stipe as described in "Figure 4" In addition, the fruiting body is the structure that is valued in the life cycle of a basidiomycete. food and has commercial value. It is a fleshy structure, commonly called a mushroom.

Figure 4: Reproductive cycle of Pleurotus ostreatus.

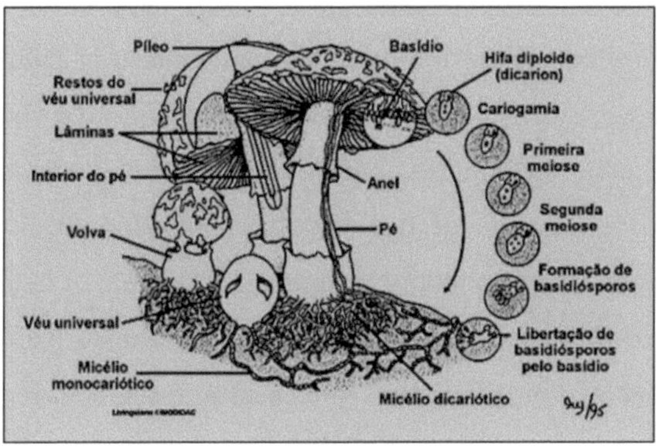

Source: LIVINGTONE (1995).

The fruiting body or mushroom of Basidiomycetes has two main functions: The first of these is the protection of the structures where karyogamy, meiosis and spores "figure: 5" are formed. Karyogamy and meiosis take place in the basidia; consequently, protecting the basidia is one of the mushroom's functions.

Figure: 5 Stages are nuclear fusion (karyogamy) and reductive nuclear division (meiosis).

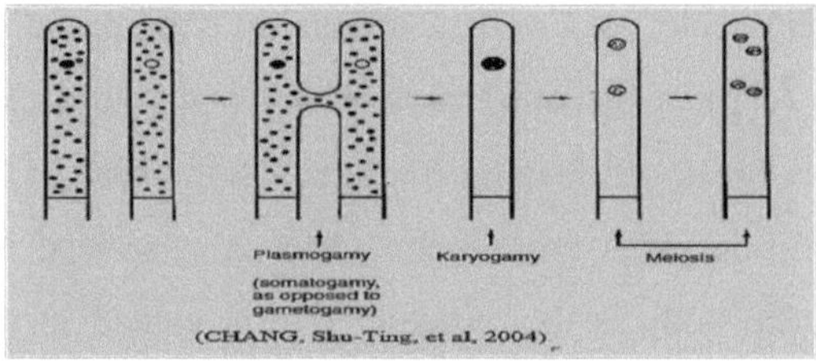

The second function of the mushroom is to provide effective dissemination of the spores that are produced in the basidia (CHANG, 2020). Although it is generally accepted that enzyme catalysis is fundamental to the development of more sustainable processes in the food and agro-industrial industries, the main focus of recent work on solid-state fermentation is less on the production of enzymes (or valuable metabolites) but on the use of the substrate to show the degradation and valorisation of lignocellulosic agro-industrial side-streams in terms of the concept of bioeconomy or simply as a source of nutrients, either for the production of fruiting bodies or for specific biotransformations (BERGER, 2022).New fermentation processes for basidiomycetes have been reported recently; one of them is the co-cultivation of two fungal strains. For example, cocultivation of Pleurotus ostreatus, the edible oyster mushroom, with Aspergillus niger (an ascomycete producer of citric acid and various enzymes) improved the degradation of aflatoxin B1, a common mycotoxin in nuts and other foods. A recent review showed how cocultures of different strains of the same or different fungal species can stimulate each other, resulting in the induction of previously silent polyketide synthase gene clusters, thus offering new options for generating bioactive polyketides. Such observations require in-depth investigation because many classic food fermentations rely on mixed cultures (BERGER, 2022). In addition to these biological characteristics, mushrooms have various nutritional properties that make them important in human nutrition. Pleurotus spp. have a high protein content, containing all the essential amino acids. This makes them an interesting alternative in vegetarian and vegan diets. Low in fat, most of which is unsaturated, which contributes to a healthy and balanced diet. They are rich in complex carbohydrates, including dietary fibre such as chitin, which helps the digestive system function and can help control blood sugar levels. Pleurotus spp. mushrooms are also significant sources of various vitamins, especially those of the B complex (such as riboflavin, niacin and pantothenic acid) and vitamin D. They may contain vitamin D2, which is important for bone health. They are rich in minerals such as potassium, phosphorus, copper, zinc and iron, contributing to electrolyte balance and cellular function, and are low in calories, making them an excellent choice for those who want a nutritious and balanced diet (KALAC, 2013).

3.3 Techniques for growing edible mushrooms

Axenic cultivation of edible mushrooms is a method that involves production under totally sterile conditions, avoiding any contamination by unwanted microorganisms. This type of cultivation is essential to guarantee the purity of mycelium growth and maximise productivity. Main characteristics of axenic cultivation: sterile environment: The process is carried out in controlled and sterilised environments, such as laminar flow chambers. Sterilisation of substrates: the substrates used in cultivation (such as sawdust, straw or agro-industrial waste) are autoclaved to eliminate any presence of contaminants. Isolation of the mycelium: the mushroom mycelium is isolated and inoculated into sterile culture media to guarantee its growth without the interference of bacteria, other fungi or pathogens. Stages of axenic cultivation: preparation of the culture medium to maintain the primary matrix: a nutrient medium is used, such as potato-dextrose agar (BDA), which is sterilised in an autoclave. Sterilisation of materials: all materials, such as flasks, tools and substrates, are autoclaved or irradiated with UV to guarantee the elimination of contaminants. Inoculation in a sterile environment: the mycelium is transferred to the culture medium in a laminar flow chamber. This ensures that the environment remains sterile during inoculation. Incubation of the mycelium: the cultures are incubated under controlled temperature and humidity conditions so that the mycelium can grow and colonise the substrate. Transfer to substrate: once the mycelium has grown in the culture medium, it is transferred to the main substrate (straw, sawdust, etc.), which has also been previously sterilised. Fruiting: When the mycelium has completely colonised the substrate, the environmental conditions are changed to induce fruiting (mushroom production). Harvesting: the mushrooms are harvested once they have reached the ideal size. The substrate can continue to produce for more than one harvest cycle, depending on the mushroom species and the technique applied. The advantage of axenic cultivation is the purity of all the material used. As the entire process is carried out under sterile conditions, the resulting mushrooms are free from contamination. High productivity: the strict control of environmental conditions promotes faster and more efficient growth of the mycelium. Food safety: The method ensures that the mushrooms are safe to eat, with no risk of contamination by pathogens (ALMEIDA et al., 2018).

The Jun-Cao technique is a method of mushroom cultivation, developed in China in 1983, which uses grasses and other herbaceous plants as a substrate instead of wood or sawdust, traditional methods of mushroom cultivation. "Jun" means mushroom and "Cao" means "mushroom". means herbaceous plant, reflecting the sustainable and efficient nature of this method. This technique is widely recognised for its environmental, social and economic benefits.

Stages of the Jun-Cao technique: selection and preparation of the substrate: the Jun-Cao technique uses grasses such as elephant grass, napier grass or agricultural crop remains. The selected plants are harvested and chopped into small pieces. The chopped material is then dried and sterilised to eliminate micro-organisms and contaminants, ensuring a favourable environment for the growth of the mycelium. Inoculation of the mycelium: the mushroom mycelium is inoculated into the sterilised substrate. This process can be carried out under sterile conditions to avoid contamination. The mycelium is first grown on agar plates or grains until it completely colonises the medium, and then transferred to the substrate. Incubation: the substrate inoculated with mycelium is placed in plastic bags or other suitable containers and kept under controlled temperature, humidity and ventilation conditions. During the incubation period, the mycelium colonises the substrate and prepares for fruiting. Fruiting: once colonisation is complete, the growing environment is adjusted to stimulate fruiting. This can involve changes in temperature and humidity. The mushrooms begin to sprout, and the fruiting process can last several weeks, allowing for multiple harvests. Harvesting: the mushrooms are harvested when they reach the right size. The Jun-Cao technique allows mushrooms to be produced quickly and efficiently, with the substrate being able to be reused on some occasions or as organic fertiliser after production has finished. Advantages of the Jun-Cao technique: sustainability. It uses agricultural waste and grasses, which reduces pressure on forest resources and promotes the efficient use renewable materials. Low cost: by using widely available materials such as grasses, the cost of production is reduced. Better productivity: the technique allows for high yields and is suitable for various species of edible and medicinal mushrooms. Environmental contribution: as well as being a solution for mushroom production, the technique helps recycle agricultural waste and contributes to reducing environmental pollution. This technique is becoming increasingly popular due to its affordability and the fact that it promotes sustainable mushroom production, especially in rural areas and regions with scarce natural resources (CHANG, 2020)

3.4 Lignocellulosic substrates used in mushroom cultivation

Various substrates are mentioned for their nutritional properties and ability to support fungal growth. Some of the most common substrates include:

1. Agricultural waste: wheat straw, rice straw and maize straw are widely used to grow mushrooms such as Pleurotus spp. and Agaricus bisporus.

2. Forest residues: the use of sawdust and bark in mushroom cultivation is well documented. Gregori et al. (2007) described the use of sawdust from various tree species in the cultivation of P. ostreatus. In addition, Bonatti et al. (2004) investigated different wood residues in mushroom cultivation.
3. Industrial waste: substrates such as coffee grounds and brewery bagasse, among others.
4. Urban waste: the use of urban waste such as cardboard or garden waste for mushroom cultivation has been explored in studies such as that by Royse et al. (2004), who discuss the potential of this waste as an alternative substrate.
5. Compost substrates: mixed composts, such as manure compost, are often used to grow Agaricus bisporus, as described by Zied et al. (2011) and Siqueira et al. (2012).

3.4.1 Bean cultivation

The bean crop (Vigna unguiculata, L.) is one of the most important in world agriculture, especially in Latin America and Africa, where it is an essential source of vegetable protein for the population. In Brazil, beans are a dietary staple and a widely cultivated crop, with significant production in states such as Paraná, Minas Gerais and Goiás (MELO, 2020). Bean production varies according to the region and the type of bean planted (carioca beans, black beans, among others). The growing cycle is relatively short, from 85 to 100 days, and the crop can be planted at various times of the year, allowing harvests in different . In Brazil, the average yield is around 1,200 kg/ha, with variations depending agricultural practices and climatic conditions Bean production generates a significant amount of waste, mainly leftovers from agricultural processing:

1 Straw: the plant residue left after the bean harvest, including stems, dried leaves and pods.

2Pods: the husks of the pods that surround the beans. After harvesting, these pods are discarded or used for alternative purposes (Figure 6).

Figure 6: Vigna unguiculata L. bean pods with grains.

Source: https://www.blogdorobertoararipina.com.br/preco-do-feijao- verde-registra-a aumento-nas-feiras-de-petrolina/feijao-verde/

3 Leftover beans that are not fit for consumption: broken, poorly formed or inadequately sized beans that are not used for direct consumption (MOTA et al., 2015).

The waste generated can be utilised:

1 Animal feed: the use of straw and pods as feed for ruminants (FERREIRA et al., 2012).

2 Organic fertiliser: composting bean waste has been evaluated by (SOUZA et al. 2002) as a sustainable strategy for improving soil fertility.

3 Biomass for energy production: the use of bean waste as a source of biomass has been studied by (RAMOS et al., 2011) in the context of renewable energies.

4 Substrate for mushroom cultivation: the use of agricultural waste in the cultivation of mushrooms, such as P. ostreatus (FIGUEIRÓ, 2011).

5 Bioplastics Manufacture of bioplastics from agricultural waste (JÚNIOR, 2020). The efficient utilisation of bean can make a significant contribution to agricultural sustainability, reducing environmental impact and promoting a

circular economy, where waste is transformed into new resources, such as mushroom production.

3.4.2 Growing algaroba

The algaroba (Prosopis juliflora) The algaroba, in addition to its adaptation to the soil and climate conditions of the Brazilian semi-arid region, offers economic potential that could make a significant contribution to the development of these regions. Native to arid and semi-arid regions of the world, the species is prevalent in almost all the states of the Brazilian Northeast. This fast-growing, drought-resistant legume with high energy potential has had numerous applications and uses in rural areas since ancient times "Figure 07" (GOUVEIA, 2023).

Figure 7: Aspect of the algaroba (Prosopis juliflora) crop.

dicasonline.com/algaroba/ onggreenday.wixsite.com/pernambuco/algaroba

The production of algaroba generates different types of waste that can be utilised in a sustainable way:

1 Pods: the pods are algaroba's main product and are used in animal feed.

However, once the pulp has been extracted, the peel and seeds are usually discarded.

2 Pruning debris: tree maintenance generates pruning debris such as branches

and leaves, which are often burnt or discarded.
3 Wood: algaroba wood is used for firewood and charcoal production, but some of the waste, such as chips and sawdust, is often not properly reused. 4 Leaves and roots: the and roots, although less exploited, are sources of raw material. (SILVA, 2009).

The waste generated by growing algaroba can be reused in various ways:

1 Animal feed: the pods are highly nutritious, rich in sugars and proteins and are widely used in ruminant, goat and pig feed. nutritional value of the pods, highlighting their importance in animal feed, especially in dry regions (MORAES, 2011).

2 Organic fertiliser: the waste from the pods, husks and leaves can be composted and used as organic fertiliser, as it improves soil fertility and promotes nutrient cycling (SIQUEIRA, 2010).
3 Energy production: wood is an important source of firewood and charcoal, especially in areas where fossil fuels are scarce. In addition, briquettes can be produced from wood waste, a more efficient alternative for energy production (NETO, 2023).
4 Substrate for growing fungi: algaroba waste, such as wood and pods, is rich in cellulose and hemicellulose and can be used as a substrate growing edible mushrooms. Queiroz et al. (2018) highlight the efficiency of this substrate in the cultivation of P. ostreatus, contributing to sustainable food production.
Utilising algaroba waste not only contributes to sustainability, but also strengthens the circular economy in arid regions. Using the plant prevents desertification, improves soil quality and provides a renewable source of energy and food. Medeiros Neto et al. (2023) also highlight the medicinal potential of the plant, with antioxidant and antimicrobial compounds present in the pods.
The efficient use of algaroba residues could be essential to promote sustainable agricultural practices, minimise waste and improve food security in arid and semi-arid regions.

3.5 Composting substrates and the impact of the carbon/nitrogen ratio on mushroom cultivation

Composting substrates is an essential stage in mushroom cultivation, as it involves preparing the substrate that will serve as a nutritional base for the fungi to develop. The composting process results in the decomposition of organic matter into a form that is more available to the mushrooms, directly influencing vegetative growth and the production fruiting bodies. A crucial factor affecting the quality of this process is the carbon:nitrogen (C:N) ratio of the substrate.

The substrates used to grow mushrooms generally include materials rich in carbon (C), such as cereal straw, sawdust and agricultural waste, and sources of nitrogen (N), such as wheat bran, soya meal or manure. These materials must be combined and composted to allow for the controlled decomposition of nutrients, promoting a suitable environment for the fungi to grow. The ideal C:N ratio in the substrate for mushroom cultivation varies depending on the species, but in the treatments of the bean substrates, Vigna unguiculata, L. and Prosopis juliflora it oscillated between 35.62:1 (T2) and 35.75:1 (T3). A proper balance of this ratio is crucial, as an excess of carbon can slow mycelium growth due to a lack of available nitrogen, resulting in lower mushroom production. Excess nitrogen, on the other hand, can promote the growth of bacteria and other undesirable microorganisms, competing with the fungi and inhibiting the growth of the mycelium. Therefore, the C:N ratio plays a fundamental role in the colonisation of the substrate by the mycelium, where an appropriate balance stimulates the efficient decomposition of cellulose and the availability of essential nutrients (CHANG, 2004). In addition to the C:N ratio, other factors affect mushroom production. Vegetative growth occurs when the mushroom mycelium colonises the substrate. The temperature, humidity and aeration of the substrate are essential for this process. Most mushrooms grow between 20°C and 30°C during mycelial growth. Higher temperatures can jeopardise development. Humidity of around 60% to 70% allows air to circulate and avoids excess water, which can jeopardise the development of the mycelium.

The fruiting phase, which is the transition from vegetative growth to fruiting, is triggered by changes in the environment, such as a decrease in temperature and an increase in humidity. The quality of the substrate influences the formation of the fruiting bodies. Well-composted substrates with a balanced C:N ratio promote higher mushroom production and quality. Adding supplementary sources of nitrogen and minerals (such as calcium, phosphorus and potassium) during composting can improve productivity. The use of soya or wheat bran during the composting phase increases the amount of protein available in the substrate, stimulating the vegetative development of the mycelium and the subsequent formation of the fruiting bodies. The pH also affects the decomposition of the compounds in the substrate and mycelial growth. A slightly acidic to neutral pH, between 6 and 7, is generally the most suitable most mushroom species. Aeration and oxygen are also important parameters in this growth and fruiting process. During the composting process and colonisation of the substrate, adequate aeration is necessary to prevent the build-up of carbon dioxide and promote aerobic respiration of the mycelium. Excess CO_2 can inhibit vegetative growth and jeopardise fruiting (ROYSE, 2017).

4 MATERIAL AND METHODS

The experiments were conducted at the Laboratory of Molecular Genetics and Plant Biotechnology (LGMBiotec/CBIOTEC/UFPB), Biotechnology Centre, Federal University of Paraíba (UFPB), Campus I, João Pessoa, Paraíba.

4.1 Origin of the Mushroom

The matrix of P. ostreatus var. florida belongs to the fungal culture collection of LGM Biotec/CBIOTEC/UFPB and is being maintained on 2 % Sabouraud-dextrose agar medium (5 g/L meat peptone, 20 g/L glucose, 5 g/L casein peptone and 15 g/L bacteriological agar) "Figure 08", according to the methodology described by Furlan et al. (1997).

Figure 8: Colony of Pleurotus ostreatus var. florida.

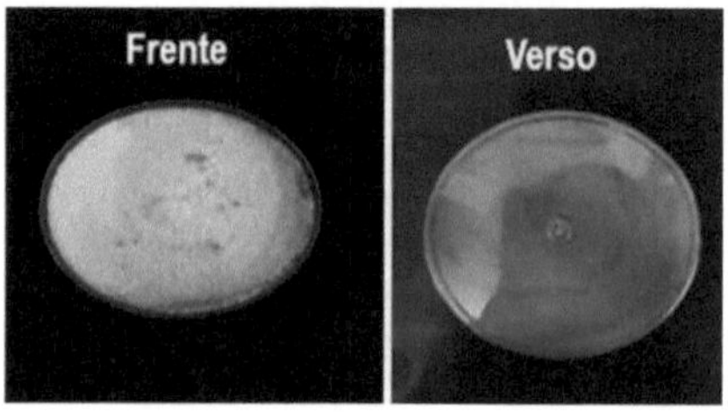

Source: Author, 2024.

4.2 Substrates used

Popcorn maize (Zea mays var. everta) of commercial origin was used, as well as green bean pods without grains (Phaseolus vulgaris) acquired from a street market in the metropolitan region João Pessoa/PB. The algaroba leaves (Prosopis juliflora (Sw) DC) were collected from Fazenda Pernambucana in São Mamede/PB "Figure 09".

Figure 9: Processed substrates for the production of Pleurotus ostreatus var. florida.

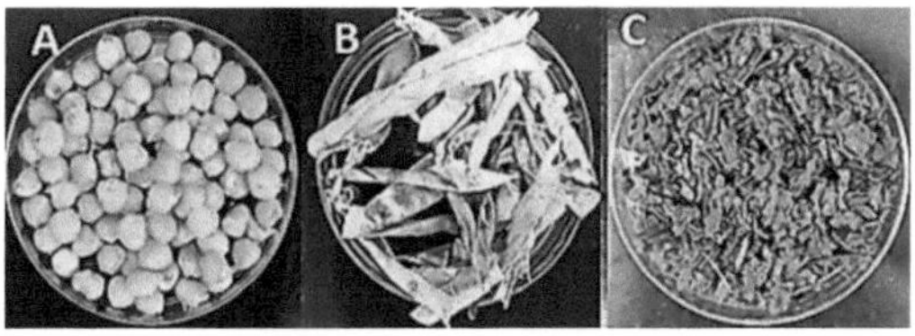

A - Popcorn corn, B - Bean pods without grains and C - algaroba leaves.
Source: Author, 2024.

4.3 Origin of substrates

The algaroba leaves were collected at Fazenda Pernambucana in São Mamede/PB. The green bean pods, after threshing the grains, were bought at a free market (Mercado Público da Torre) in the metropolitan region of João Pessoa/PB. The popcorn used was of commercial origin.

4.4 Substrate Processing

Popcorn (Yoki): the kernels were cooked over high heat in a pressure cooker for 90 minutes. After cooking, the extract from the cooking process was drained and the kernels washed to remove excess starch and left at room temperature (28± 2°C) to dry externally for approximately 60 minutes. They were then packed in plastic bags and kept refrigerated at 4°C (BONATTI et al., 2004).

Bean pods without the grain: initially they were dehydrated at 70ºC for 72 hours in a drying oven. Next, it was submerged in tap water for 24 hours and, after draining the water, it was placed on a flat platform for 2 hours to dry. Finally, the pods were ground in a knife mill and packed in plastic bags and kept refrigerated at 4°C until use.

Algaroba leaves: after collection, the leaves were placed in a drying oven for 72 hours at 60 ± 5^0C. After drying, the leaves were submerged in running water for

24 hours. The leaves were then pasteurised by immersing them in boiling water for 3 hours. Afterwards, the water was drained off and the leaves were left on a solid platform to dry externally for approximately 2 hours at room temperature (28± 2^0C). They were then packed in plastic bags and kept refrigerated at 4^0C.

4.5 Seed production

Initially, the processed popcorn was placed in a glass jar (6.0 x 8.5 cm) filling 80% of its volume. The jar was then closed with aluminium foil and autoclaved at 121°C for 15 minutes. After cooling to room temperature (28° ± 2°C), three discs of mycelium with a diameter of 5 ± 1 mm were inoculated into the flask. Incubation took place at 28° ± 2°C, in the absence of light, until the surface was completely colonised (Figure 10).

Figure 10: Seed or spawn: maize colonised with Pleurotus ostreatus var. Florida.

Source: Author, 2024.

4.6 Qualitative assessment of mycelial growth

The substrates were weighed and mixed (final weight 60 g) (Table 1). The treatments were placed in polythene plastic bags (2.5 x 15 cm), partially closed with polyamide ties and autoclaved at 121°C for 60 minutes (Figure 07). After 24 hours, 3 corn kernels with adhered mycelium were added to each bag. The treatments were incubated at 28° ± 2°C in the absence of light for 15 days. Longitudinal growth was measured with a millimetre ruler (cm) to obtain the average mycelial growth of each formulation. The total colonisation time of each treatment was assessed, taking into account the time needed for all the specimens in the triplicate to reach complete colonisation of the surface. Popcorn was used as a control. To assess the vigour/density of the mycelium during the mycelial run, the treatments were compared qualitatively (XIEMENES, 2019).

Table 1: Formulations for growing Pleurotus ostreatus var. florida.

Treatments	Formulations/proportions
TC - control	Popcorn
T1	Green bean pods without beans (95%)+ algaroba leaves (5%)
T2	Green bean pods without beans (90%)+ algaroba leaves (10%)
T3	Green bean pods without beans (80%)+ algaroba leaves (20%)
T4	Green bean pods without beans (70%)+ algaroba leaves (30%)
T5	Green bean pods without beans (60%)+ algaroba leaves (40%)
T6	Green bean pods without beans (50%)+ algaroba leaves (50%)

Source: Author, 2024.

Figure 11: Formulated substrates for the production of Pleurotus ostreatus var. florida.

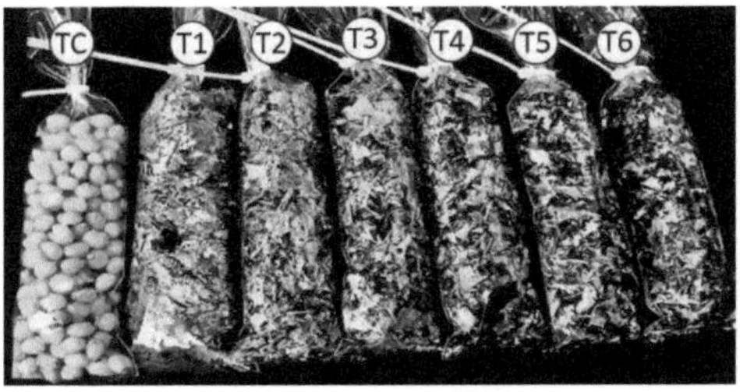

TC [Popcorn corn], T1 [Green bean pods without grains (95%)+ algaroba leaves (5%)], T2 [Green bean pods without grains (90%)+ algaroba leaves 10%)], T3 [Green bean pods without grains (80%) + algaroba leaves (20%)], T4 [Green bean pods without grains (70%) + algaroba leaves (30%)], T5 [Green bean pods without grains (60%) + algaroba leaves (40%)] and T6 [Green bean pods without grains (60%) + algaroba leaves (40%)]. Source: Author, 2024.

4.7 Formulations for production

For this experiment, 100g of treatments T1, T2, T3, T4 and T6 were weighed according to the proportions described in Table 01. The treatments were placed in polythene plastic bags (12 x 20 cm), partially closed with polyamide ties and autoclaved at 121ºC for 60 minutes. After 24 hours, 3 corn kernels with adhered mycelium were added to each bag. The treatments were incubated at 28° ± 2°C in the absence of light for 15 days. The experiment was carried out in quintuplicate. After incubation, the bags were opened to assess the fruiting potential. The fruiting bodies were harvested when they showed maximum development, as verified by the beginning of the unwinding of the pileus margins (XIEMENES, 2019).

4.8 Evaluation of biological parameters

During the cultivation period, biological aspects of the mushroom were analysed, such as: colonisation time of the substrate, primordia emission period, basidiomata formation time, stipe height and width (cm) and pileus diameter (cm) (XIEMENES, 2019).

4.9 Evaluation of production parameters

Yield, biological efficiency, loss of organic matter, compost consumption, moisture and productivity were assessed.

Yield: measured in terms of the fresh weight (g) of mushrooms produced per bag containing 100g of compost (ZAKIL et al., 2022).

Biological efficiency (EB%): the biological efficiency (EB%) of the process was determined by the ratio between the mass of fresh fruiting bodies and the mass of substrate based on their dry weight according to Equation 01 (MUSWATI et al., 2021).

$$EB(\%) = \frac{MBS}{MSS} \times 100 \quad \text{Eq. 01}$$

MBS= Mass of fresh basidiocarps, g MSS = Dry mass of substrates, g

Productivity: the productivity (Pr g/day^{-1}) of the process was determined according to Holtz (2009). It consists of the ratio between the mass of dried fruiting bodies and the total time time from inoculation to production flow, as shown in Equation 02.

$$Pr(g/dia) = \frac{MCS}{TC} \quad \text{Eq. 02}$$

Where: MCS= Dry mass of fruiting bodies, g; TC= Total cultivation time, days.

Loss of organic matter: an index that evaluates the decomposition of the substrate by the fungus. This index is based on the loss of organic matter decomposed by the fungus, which is determined by the difference between the dry mass of the initial substrate and the dry mass of the residual substrate (Equation 03) (RAJARATHNAM; BANO, 1989).

$$PMO\,(\%) = \frac{(MSSI - MSSR) \times 100}{MSSI} \quad \text{Eq. 03}$$

In which:

PMO= Organic matter loss, % MSSR= Residual substrate dry mass, g MSSI= Initial substrate dry mass, g

Compost consumption percentages: this index was calculated based on the dry mass of the substrate after cultivation divided by the initial dry mass of the compost, as shown in Equation 04.

$$CC(\%) = \frac{CF \times 100}{CI} \quad \text{Eq. 04}$$

Where: CF= Final dry matter compost; CI= Initial dry matter compost.

Moisture: this was determined by the gravimetric method using heat, which is based on the weight loss of the material when heated until it reaches a constant weight. All the treatments were kept in a drying oven at 105 ºC for 24 hours and the moisture content of the substrate was determined by the difference between the wet and dry biomass (Equation 05).

$$U(\%) = \frac{(MI - MF) \times 100}{MI} \quad \text{Eq. 05}$$

Where: MI= Initial sample mass; MF= Final sample mass.

4.10 C, N and C:N content

Organic carbon: determined using the Walkley and Black method, modified from Tedesco et al. (1995). Each sample (0.01 g) was added to a flask containing potassium dichromate ($K_2Cr_2O_7$) and sulphuric acid (H2SO4) and heated to 60 ºC. Two types of white were prepared: one that was subjected to the same heating conditions and one that was not heated. After cooling, indicators were added and then titration was carried out with ammoniacal ferrous sulphate solution (Fe (NH4)2(SO4)2.6H2O). The results were obtained from Equations 06 and 07:

$$A = \frac{((Vba - Vbn)(Vbn - Vba))}{Vbn + (Vba - Vam)} \quad \text{Eq. 06}$$

In which:

Vba: volume spent titrating the blank with heating; Vbn: volume spent titrating the blank without heating; Vam: volume spent titrating the sample.

$$Carbono\ (C)\% = \frac{(A)(molaridade\ do\ sulfato\ ferroso)(3)\ x\ 100}{(massa\ da\ amostra\ (mg))} \quad \text{Eq. 07}$$

In which:
3= ratio between the number of moles of dichromate reacting with iron, multiplied by the number of moles of dichromate reacting with carbon, multiplied by the atomic mass of carbon.

Total nitrogen: the Kjeldahl method was used to analyse total nitrogen, where the samples (0.1g) were digested in a catalyst mixture containing sulphuric acid (H2SO4) in a Velp Scientifica digester block until a blue colour was obtained. After cooling, the reactions were distilled in a Velp Scientifica nitrogen distiller in a basic medium (NaOH 35%) and the product of this distillation was collected in an erlenmeyer flask in a 4% boric acid solution (H3BO3) with indicators and titrated with 0.1 N hydrochloric acid (HCl) (Tedesco et al., 1995, Moretto et al., 2002a). The nitrogen concentration was determined according to Equation 08:

$$N(N)\% = \frac{V\ x\ Fc\ x\ 0{,}0014}{M} \quad \text{Eq. 08}$$

In which:
V= volume of HCl used in the titration, mL. Fc= HCl correction factor
M = sample mass (g)
C:N ratio: The C/N ratio was obtained from the proportion of carbon contained in each sample in relation to nitrogen.

4.7 Statistical analysis

The experimental design was entirely randomised complete blocks, using the Tukey test, with a $p<0.05$ significance level (GOMES, 1990). The statistical analyses were carried out using SAS (Statistical Analysis System) developed by

SAS Institute Incorporation. The results were analysed using Tukey's test with a 5% significance level (ANOVA).

This test compares all the means to determine whether there are significant differences between the treatments and that differences observed with a probability of less than 5% are considered statistically significant by the analysis of variance (ANOVA); In the experimental design: They were randomised to ensure that all units had the same chance of receiving any treatment. This helps to ensure that the differences observed in the results are due to the treatments applied and not to external or systematic factors (ANOVA).

5 RESULTS AND DISCUSSION

5.1 Biological cycle of Pleurotus ostreatus var. florida

The stages of mushroom production in artificial environments basically represent their life cycle, with evidence of their macro and microscopic morphological characteristics and their vegetative and reproductive aspects (Figure 12). Figure 12A shows the appearance of the primary colony/matrix of P. ostreatus var. florida on Sabouraud-dextrose agar culture medium after 15 days growth. The white, dense appearance is very characteristic of the species (Figure 12B1), the connecting staple (Figure 12B2) and the septum are structures that connect hyphae, common in basidiomycete fungi (Figure 12C). Seed production (inoculum) on pre-cooked maize grains as a substrate (Figure 12C). Seed production is a crucial stage in production. Maize serves as a nutrient base for the initial growth of the fungus, facilitating its inoculation into production substrates. After inoculating the seeds, the fungus grows vegetatively for 15 days on specific substrates (green bean pods (95%) and algaroba leaves (5%)) (Figure 12D). The fruiting phase is characterised by the formation of the mushroom or pileus/hat/basidiome (Figure 12E1), which is the edible part of the fungus. In the lower part of the basidiom we can see the stem (stipe), the structure of the mushroom's hymenium, which has a lamellar shape (Figure 12E2) where the spores are produced. Below is the base or foot (Figure 12E3), where the stipe that supports the pileus is attached to a base or foot (the lower part of the stem that anchors the mushroom to the substrate). Lamellar-shaped myrmenium in (Figure F) and the sexual spores (Figure G).

Figure 12: Vegetative and reproductive aspects of Pleurotus ostreatus var. florida.

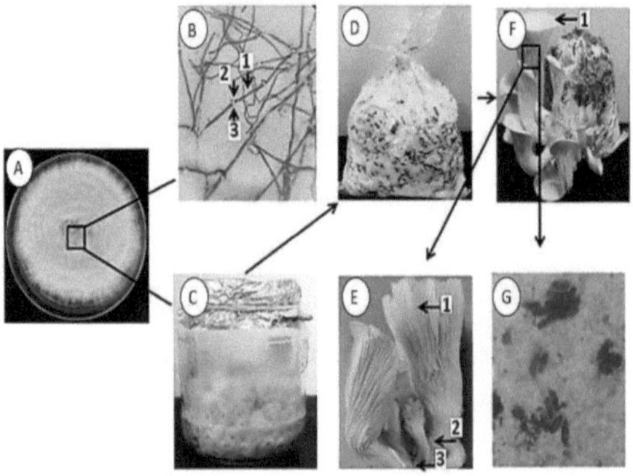

A: Colony of Pleurotus ostreatus var. florida. with 15 days of growth on Sabouraud-dextrose agar culture medium); **B**: Vegetative aspect: 1 - hyphae, 2 - connecting staple and 3 - septum; **C**: production of Pleurotus ostreatus var. florida seed on pre-cooked maize grain; **D**: Vegetative growth (green bean pods (95 %) + algaroba leaves (5.0 %) in production bags 15 days after inoculation; **E**: Fruiting, 1 - Mushroom or hat or basidiom, 2 stem, 3 base or foot; **F**: 1 - Lamellar-shaped hymenium; **G**: Sexual spores. Source: Author, 2024.

5.2 Evaluation of the mycelial development of Pleurotus ostreatus var. florida in different formulations

When characterising growth, total mycelial development time varied from 15 days (TC) to 23 days (T1, T2, T3), with T4, T5 and T6¨ 22, 20 and 20, respectively. Mycelial development in TC reached 11.50 cm, significantly higher than all the other treatments. Among the treatments, T4 had the highest mycelial growth (9.40 cm), followed by T1 (9.10 cm) (Figure 09). In terms of mycelial growth (cm/day), TC (0.76 cm/day) showed the best performance compared to the other groups, but its growth was insignificant compared to treatments T4 (0.62 cm/day) and T2 (0.56 cm/day). The humidity of the treatments ranged from 58% to 71%. All the treatments showed abundant, continuous, dense and vigorous growth, similar to the TC group. These characteristics are essential for good fruiting (Table 02). The literature suggests that the humidity of the substrate is a critical factor for mycelial growth, as it directly affects the availability of water, which is essential for the transport of nutrients and the cellular metabolism of the mycelium (LAVELY, 2023). In this study, humidity was within the established standard (50% and 75%) and may have contributed to continuous growth as observed in all treatments. Adequate humidity ensures that the mycelium has access to sufficient water without the substrate becoming anaerobic, which could jeopardise growth. By correlating the performance of the 6 treatments with maize grain already used (TC - control group), it was possible to infer that TC, possibly due to its balanced composition and humidity, provided a more favourable environment for the rapid and efficient growth of the mycelium as expected. Although mycelial growth in terms of cm/day was higher in the TC (0.76 cm/day). This suggests that although TC promoted slightly faster growth, other factors such as substrate uniformity and nutrient availability also play important roles in determining the success of mycelial development (LAVELY, 2023). Humidity above 70 per cent enables the development of diseases and competing fungi. Water contaminated with heavy metals such as mercury, lead and copper can cause an undesirable flavour to the product and be a source of human contamination (BELLETINI et al., 2019).

Table 02: Evaluation of density, vigour and growth in the different substrates for the production of Pleurotus ostreatus var. florida .

Treatments	Total mycelial development time (days)	mycelial development (cm)•	Mycelial growth (cm/day)	Moisture content of formulations (%)
TC (control)	15,00 b	11.50 cm	0,76 a	58%
T1	23,00 a	9,10 cm	0.60 ab	71%
T2	23,00 a	8.50 cm	0,56 b	70%
T3	23,00 a	8.60 cm	0,57 b	69%
T4	22,00 a	9.40 cm	0.62 ab	68%
T5	20.00 ab	8.70 cm	0,58 b	68%
T6	20.00 ab	8.80 cm	0,58 b	68%
CV%§	2,79	4,04	3,14	

Averages followed by the same letter do not differ by Duncan's multiple range test (P<0.05). Data not transformed.*U% Moisture of the formulations. • Average growth up to the 15th day. T1 to T6 days. § Coefficient of variation.

Figure 13: Macroscopic aspect of the colonisation of Pleurotus ostreatus var. florida in the different treatments 15 days after inoculation.

Source: Author, 2024.

5.3 Characterisation of the biological and production parameters of Pleurotus ostreatus var.florida

All the treatments showed abundant development (+++). T1 to T5 were more efficient, completing the cycle in 21 days compared to T6 (23 days) (Table 03). Stipe size ranged from 1.50cm to 2.00cm. Width was most prominent in T5 (1.0 cm), followed by T4 (0.77 cm), T1 and T3 (0.63 cm), T2 (0.57 cm) and T6 (0.56

cm). In With regard to the area of the pileus, T1 formed 9.23 cm^2 of area, T2 12.74 cm^2, T3 14.31 $cm^{(2)}$, T4 10.74 cm^2, T5 10.57 cm^2 and T6 6.92 $cm^{(2)}$) (Figure 13).

Figure 14: Pleurotus ostreatus var. florida fruiting bodies produced in the different treatments.

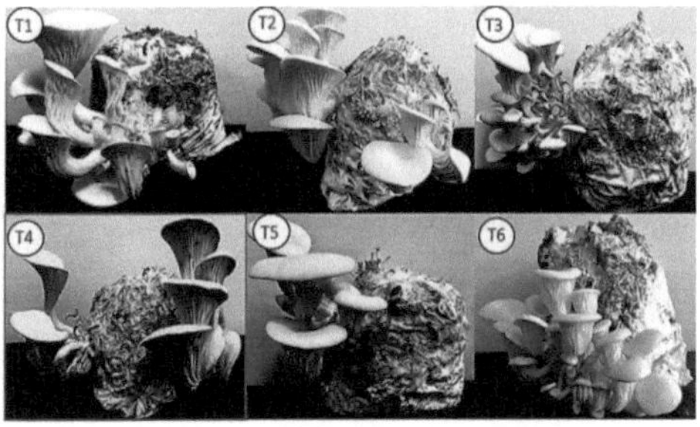

Source: Author, 2024.

Table 03: Biological parameters of Pleurotus ostreatus var. florida found in the different formulations

Treatments	I developed -mycelial growth	Vegetative growth (days)	Total cycle (days)	Stipe size (cm)	Stipe width (cm)	Pileus area (cm2)
T1	+++	15	21	1,67±0,17	0,63±0,92	9,23±0,47
T2	+++	15	21	2,00±0,04	0,57±0,67	12,74±0,81
T3	+++	15	21	1,83±0,49	0,63±0,23	14,31±0,79
T4	+++	15	21	1,50±0,12	0,77±0,45	10,74±0,29
T5	+++	15	21	1,83±0,21	1,00±0,87	10,57±0,89
T6	+++	17	23	1,67±0,09	0,56±0,29	6,92±0,33
(±)EP				3,73	5,20	4,53
CV (%)				5,87	7,41	5,21

+ sparse growth; ++ moderate growth; +++ abundant growth. CV: Coefficient of variation. SE: Standard error. T1: Green bean pods without grains (95%)+ algaroba leaves (0.5%); T2: Green bean pods without grains (90%) + algaroba leaves (10%); T3: Green bean pods without grains (80%) + algaroba leaves (20%); T4: Green bean pods without grains (70%)+ algaroba leaves 30%); T5: Green bean pods without grains (60%). + algaroba leaves (40%) and T6 Green bean pods without beans (50%) + algaroba leaves (50%).

All the treatments formed fruiting bodies/basidioma. In the evaluation of the production parameters "Table 04" after harvest, the basidiom unit varied from 89 to 91 per cent between the treatments, indicating that the formulations used contained adequate moisture levels for the growth of P. ostreatus var. florida. The yield was relatively similar between all the treatments and ranged from 12.50 (T6) to 18.00% (T4). Biological efficiency was T1 (49.42%), T2 (52.96%), T3 (52.68%), T4 (56.25%), T5 (42.50%) and T6 (39,06%). This reflects a good conversion of nutrients from the substrate into mushroom biomass. Productivity was highest in treatments T1 (0.63 g/day), T2 (0.61 g/day) and T3 (0.61 g/day). The loss of organic matter (LOM) indicates that the substrate is being efficiently utilised to produce biomass gradually. This parameter varied from 42.52% (T1), 42.22% (T2), 48.74% (T3), 36.71% (T4), 28.75% T5) and 46.87% (T6). This This process occurs during the growth and development of the fungi, which use the nutrients in the substrate to form biomass. Compost consumption varied from 51.25% to 71.25% between all treatments.

Table 04: Production parameters of Pleurotus ostreatus var. florida found in the different formulations.

Treatments	Basidiom humidity (%)	Yield (%)	Biological efficiency (%)	Productivity (g/day)	Loss organic matter (%)	Compost consumption (%)
T1	89.00± 0.19	14,33±0,18	49,42±0,75	0,63±0,05	42,52±0,97	57,47± 0,69
T2	90,00±0,81	15,88±0,37	52,96±0,49	0,61±0,98	42,22±0,11	57,77±039
T3	91,00±0,31	16,33±1,76	52,68±1,32	0,61±0,87	48,74±0,49	51,25±0,76
T4	89,00±0,29	18,00±0,09	56,25±1,98	0,36±0,69	36,71±0,20	63,28±0,49
T5	90,00±0,48	13,60±0,71	42,50±0,98	0,29±0,16	28,75±0,09	71,25±0,72
T6	89,00±0,11	12,50±1,09	39,06±1,86	0,22±1,69	46,87±0,20	53,12±0,14
Standard deviation	4,76	8,03	2,74	5,98	3,89	6,43
CV (%)§	10,67	16,43	20,61	19,72	8,19	16,74

Untransformed data. § Coefficient of variation.

Basidiom moisture is a crucial factor that directly influences the quality and commercial acceptability of mushrooms. Recent studies indicate that high levels of moisture are desirable as they improve the texture and appearance of the final product, factors that are critical for market acceptability (ZHENG et al., 2023). However, excess moisture can reduce the shelf life of fresh mushrooms, increasing the risk of microbial contamination. Their results indicate that all treatments kept moisture within a relatively controlled range, which is positive for commercialisation. Yield is one of the main indicators of substrate use efficiency and is directly linked to sustainability, as a higher yield indicates a more efficient use of resources (BETTIOL et al., 2023). Yields of over 15 per cent are indicative of good conversion of the substrate into fungal biomass, suggesting that T4 was the most efficient in this respect. Comparison with other authors reveals that yields in this range are in line with the best sustainable cultivation practices. Biological efficiency (BE) is an essential metric for assessing the productivity and sustainability of production, representing the percentage of substrate converted into biomass. In the study by Wurmstich et al. (2021), it is reported that an EB of more than 50% is considered excellent, indicating a good adaptation of the fungus to the substrate used. Treatments T2, T3 and T4 stood out, surpassing this mark. Daily productivity is an important indicator not only of efficiency, but also of the economic viability of cultivation. Although T4 had the highest EB, its daily productivity was lower, which may indicate a slower growth rate, but one that is sustained over time. Studies such as that by Campos et al. (2020) suggest that lower daily productivity can be compensated for by greater efficiency over the cultivation cycle, as long as the quality of the end product is maintained. Organic matter loss (OML) is directly associated with substrate decomposition and the efficiency of conversion into biomass. Lower losses of organic matter are indicative of more efficient decomposition, resulting in a lower environmental impact (ZHENG et al., 2023). This condition favours more efficient decomposition, which is a positive point from a sustainability point of view. Compost consumption reflects the fungus's ability to utilise the available nutrients. Greater consumption indicates better adaptation of the fungus to the substrate and greater efficiency in biomass production. The data presented indicates that T4 and T5 had relatively high compost consumption. Higher consumption has been correlated with greater efficiency in converting substrate into biomass (ZHANG et al., 2021). Based on the parameters of yield and biological efficiency, T1, T2, T3 and T4 are promising for the sustainable production of P. ostreatus var. florida.

5.4 Chemical characterisation the formulations: carbon (C) and nitrogen (N) content and C:N ratio.

From the chemical evaluation of the substrates, the carbon varied slightly between treatments. The highest values were observed in T3 (57.2%) and the lowest in T5 (56.0%). All the treatments have relatively close carbon contents, indicating a consistency in the composition of the substrates (Table 05).

The nitrogen content ranged from 1.6% to 1.9%, with T6 having the highest nitrogen (1.9%) and T2, T3 and T5 having the lowest (1.6%). Adequate levels of nitrogen are essential for the growth and development of fungi, as nitrogen is a fundamental component for the synthesis of proteins and other biomolecules.

In terms of the Carbon:Nitrogen (C:N) ratio, treatments T2 and T3 have the highest ratios at 35.62:1 and 35.75:1 respectively. Treatment T6 had the lowest C:N ratio (29.63:1).

Table 05: Carbon and nitrogen content and :nitrogen ratio of the different formulations for growing Pleurotus ostreatus var. florida.

Treatments	C (%)	N (%)	C:N
T1	56,6	1,7	33,29 : 1
T2	57,0	1,6	35,62 : 1
T3	57,2	1,6	35,75 : 1
T4	56,7	1,7	33,35 : 1
T5	56,0	1,6	35,00 : 1
T6	56,3	1,9	29,63 : 1

C(%): Carbon; N(%): Nitrogen; C:N: Carbon to Nitrogen ratio.

The presence of carbon (C) in the substrate acts as the main source of energy for fungi. It is used in respiration to produce ATP, which is necessary for all metabolic activities. In addition, carbon is a fundamental structural component of fungal cells. Carbon-rich substrates provide the energy needed for mycelium expansion. An adequate supply of carbon is essential for cell proliferation and the formation of a dense and healthy mycelial network, and carbon is used to build the structures of the fungi. fruiting bodies (mushrooms). Therefore, substrates with sufficient carbon content are crucial for the development of large, well-formed mushrooms (LAVELY, 2023). The presence of nitrogen (N) in the substrate is essential for the synthesis of amino acids, which are the building blocks of proteins. Proteins, including enzymes, are necessary for practically all metabolic processes. Substrates with adequate nitrogen content promote healthy mycelium growth, as nitrogen is necessary for the production of structural and functional proteins. Nitrogen is also needed during the fruiting phase, as it is used in the production of enzymes and other proteins that help in the formation and maturation of fruiting bodies (LAVELY, 2023). P. ostreatus var. florida is known for its ability to degrade a wide range of lignocellulosic substrates. Its ability to utilise different sources of carbon and nitrogen allows it to be cultivated in diverse environmental conditions and with different types of agricultural waste. The choice and formulation of the substrate must take into account the appropriate C:N ratio (SILVA et al., 2024). Waste rich in carbon (algaroba) and waste rich in nitrogen (bean husks) were used as substrates. The right balance of these nutrients in the substrate was essential to promote vigorous mycelial growth and abundant production of fruiting bodies. Understanding and manipulating these nutritional factors could lead to greater efficiency and productivity in mushroom cultivation.

6 CONCLUSIONS

Based on the results obtained under the conditions established and evaluated in this study, it was possible to conclude:

- The use of bean pods without grains and algaroba leaves proved promising in the formulation of a substrate for mushroom cultivation and production, as they supported vegetative development and basidiom formation;

- The total cycle in the six treatments was relatively short, from 21 to 23 days, and characteristic basidiomes were found with the expected size for the P. ostreatus var. florida species;

- Based on the production parameters (yields and biological efficiency), formulations T1, T2, T3 and T4 were more promising, as they showed better conditions for vigorous mycelial growth and fruiting;

- All treatments showed a C:N ratio within the range considered good for the vegetative and productive development of P. ostreatus var. florida.

REFERENCES BIBLIOGRAPHIC

ALMEIDA, Arthur Costa Pereira Santiago de; SILVA, Laís Marinho de Melo Marques da; BRITO NETO, José Soares; et al. Axenic cultivation of edible mushrooms in agro-industrial waste. in: I regional meeting of agro-environmental studies: socio-environmental responsibility of scientific research, 3., 2018, Rio Largo. Revista Craibeiras de Agroecologia, Rio Largo, v. 3, n. 1, p. e6651, 2018.

AMAZONAS, Angela. Report on mushroom production in Brazil. São Paulo: National Association of Mushroom Producers, 2013.

ANDRADE, M. C. N. Applicability of the use of waste from different banana cultivars

ANPC (National Association of Mushroom Producers). Report on Mushroom Production in Brazil - 2020. Brasilia, 2020.

BELLETINI, Marcelo Barba; fiorda, Fernanda Assumpção; et al, Analysis of factors affecting the mushroom Pleurotus spp. Saudi Journal of Biological Sciences, v. 26, p. 633-646, 2019. Produced and hosted by Elsevier BV on behalf of King Saud University.

BERGER, r. g.; ERSOY, f. Improved foods using enzymes from basidiomycetes.

Processes. MDPI (Multidisciplinary Digital Publishing Institute), 2022.

BETTIOL, Wagner; SILVA, Carlos Alberto; et al. (Ed.). Understanding soil organic matter in tropical and subtropical environments. Brasília, DF: Embrapa, 2023.

BONATTI, M. et al. Evaluation of Pleurotus ostreatus and Pleurotus sajor-caju nutritional characteristics when cultivated in different lignocellulosic wastes. Food Chemistry, v. 88, n. 3,
p. 425-428, 2004. Available at: https://doi.org/10.1016/j.foodchem.2004.01.050. Accessed on: Aug. 2023.

CAMPOS, Thiago Souza; SOUSA, westefann dos santos; OLIVEIRA junior, Valdivino Domingos de. Use of biostimulants to increase grain productivity. Revista Agrotecnologia, Ipameri, v. 11, n. 1, p. 9-15, 2020.

CAPRA, r.s.; tonin, f.b.Rise of edible mushroom cultivation in Brazil. in: jornada científica e tecnológica da fatec de botucatu, Botucatu. Proceedings. São Paulo: Fatec, 2019.

CARVALHO, c.; DE AGUIAR, l. v.; SALES-CAMPOS, c.; minhoni, m. t. a.; de CHANG, r.; mushi, e. m.; he, q. Jun-Cao technology: Advances in the cultivation of edible and medicinal mushrooms for sustainable agricultural development. Agricultural Sciences, v. 11, p. 234-243, 2020.

CHANG, Shu-Ting; MILEs, philip g. Mushrooms: Cultivation, Nutritional Value, Medicinal Effect, and Environmental Impact. 2. ed. Boca Raton: CRC Press, 2004.

FAN, L. et al. Cultivation of Pleurotus mushrooms on Brazilian coffee husk and effects of caffeine and tannic acid. Mycologia Aplicada International, v.15, p.15-21, 2003.

FAO - Food and Agriculture Organisation. FAOSTAT: Agricultural production statistics. Available at: https://www.fao.org/faostat/en/#data/QV. Accessed on: 19 October 2024.

FERREIRA, a.l.; maurício, r.m.; pereira, l.g.r.; azevêdo, j.a.g.; oliveira, l.s.; pereira, j.m. Nutritional divergence in genotypes of forage peanut. Revista Brasileira de Zootecnia, v.41, n.4, p.856-863, 2012.

FIGUEIRÓ, Gláucia Garcia; GRACIOLLI, Luiz Antônio. Influence of the chemical composition of the substrate in the cultivation of Pleurotus florida. Ciência e Agrotecnologia, Lavras, v. 35, n. 5, p. 924-930, Sep/Oct 2011.

FURLAN, S. A. et al. Mushroom strains able to grow at high temperatures and low pH values. World Journal of Microbiology and Biotechnology, vol. 13, p. 689-692, 1997.

GOMES, F. P. Curso de Estatística Experimental. Piracicaba: Nobel, ed.13, 1990, p. 476, 1990. GOMES, M. National Association of Mushroom Producers

(ANPC). Agribusiness: Consumption and production of mushrooms grows in Brazil. 29/01/2018.

GOUVEIA, Clovis. Algarobeira. Laboratory for Analyses and Research into Alcoholic Beverages, Department of Chemical Engineering, CT, UFPB, 2023.

GUARÇONI, a.; FAVARATO, l. f.; STIPP, s. r.; CASARIN, v. Soil fertility management for more sustainable agricultural production. Incaper em Revista, Vitória, v. 10, p. 22-42, 2019.

HOLTZ, M. et al. cultivation of Pleurotus ostreatus using cotton waste from the textile industry. Journal of Environmental Sciences, v. 3, n. 1, p. 37-51, 2009. Available at: https://www.researchgate.net/publication/279664685_cultivo_de_Pleurotus_ostreatus_utilizan do_residuos_de_algodao_da_industria_textil. Accessed on: Aug. 2023.

JANG, myoung-jun; LEE, yun-hae. Effect of nutrient substrates on Pleurotus ostreatus in bottle culture. Environmental Agriculture Research Division, GARES, Hwaseong, Gyeonggi-Do, 445-784, Korea; Mushroom Research Institute. 2021.

JÚNIOR Santana, Cláudio Carneiro. Using lignocellulosic biomass from the Legal Amazon to produce bioproducts in an economic and socio-environmental context, Palmas, 2020.

KALAC, P. A review of chemical composition and nutritional value of wild-growing and cultivated mushrooms. Journal of the Science Food and Agriculture, v. 93, n. 2, p. 209-218, 2013.

LAVELY, T. The Grower's Guide to Mushroom Substrates, MycoHaus - Gourmet Mushrooms & Cultivation Supplies. 2023.

MACHADO, A. E. V. Integrated cultivation of the Pleurotus ostreatus mushroom and tomato (Solanum lycopersicum). AGRIS - International System for Agricultural Science and Technology, 2020. MELO, L. C.; PEREIRA, H. S.; Faria, L. C.; WENDLAND, . Bean cultivation: challenges and opportunities. Santo Antônio de Goiás: Embrapa Rice and Beans, 2020.

MENDES, J. S.; GONÇALVES, M. R. A. Competitiveness Analysis of the Mushroom Market in Brazil: Challenges and Opportunities. Revista Brasileira de Agricultura, v. 23, n. 2,p. 123-135. 2021.

MORAES, Salete Alves de; COSTA, Samir Augusto Pinheiro; Araújo, GHERMAN Garcia Leal de. Nutrition and nutritional requirements. Goat and sheep production in the semi-arid region. Petrolina: Embrapa Semiárido, 2011.

MORETTO, E; Fett, R; GONZAGA, L.V; Kuskoski, E.M. 2002. Centesimal composition of plants and other materials. 2nd ed. UFRGS, 174 p. Food products. Chapter 1. in: Moretto et al. (eds). Introduction to food science.

MOTA, Diego Henrique da. Physical properties during drying and technological quality of carioca bean cultivars under different storage conditions. University of Brasília, Brasília, 2015.

MUSWATI, C., SIMANGO, K., Tapfumaneyi, L., MUTETWA, M., Ngezimana, W. The Effects of Different Substrate Combinations on Growth and Yield of Oyster Mushroom (Pleurotus ostreatus). International Journal of Agronomy, vol 21, p-10, 2021. Available at: https://www.hindawi.com/journals/ija/2021/9962285/. Accessed on: Mar. 2024.

MUSWATI, C.; Simango, K.; Tapfumaneyi, L.; Mutetwa, M.; Ngezimana, W. The effects of different substrate combinations on growth and yield of oyster mushroom (Pleurotus ostreatus). International Journal of Agronomy, v. 2021, 2021.

NAKAZAWA, Takehito; KAWAUCHI, Moriyuki; OTSUKA, Yuitsu; Han, Junxian; KOSHI, Daishiro; SCHIPHOF, Kim; RAMÍREZ, Lucia; Pisabarro, ANTONIO G.; HONDA, Yoichi. Pleurotus ostreatus as a model mushroom in genetics, cell biology, and material sciences. Mini-Review. Applied Microbiology and Biotechnology, v. 108, p. 217, 2024.

NETO Medeiros, Pedro Nicó de; MARQUES Cirilo, NAUAN ribeiro; RIBEIRO, Libânia da Silva; SILVA, João Gabriel Missia da; PAIVA, William de; ASSIS do Carmo, Flavio Cipriano de. Energy potential of algaroba wood for the ceramics industry in north-eastern Brazil. Revista em Agronegócio e Meio

Ambiente - RAMA, 2023.

PANDITA, Deepu; PANDITA, Anu. Mushrooms: nutraceuticals and functional foods. Boca Raton: CRC Press, 2023.

RAJARATHNAM, S.; BANO, Z.; STEINKRAUS, K. H. Pleurotus mushrooms. Part III. Biotransformations of natural lignocellulosic wastes: commercial applications and implications. Critical Reviews in Food Science & Nutrition, v. 28, n. 1, p. 31-113, 1989. Available at: https://doi.org/10.1080/10408398909527491. Accessed on: Aug. 2023.

RAMOS, Luana Elis de; TRUGHILO, Paulo Fernando; NAPOLI, Alfredo; BIANCHI, Maria Lúcia. Characterisation of vegetable biomass waste for use in energy generation. Cerne, Lavras, v. 17, n. 2, p. 237-246, Apr./Jun. 2011.

ROYSE, d. j.; BAARS, J. j. p.; Tan, Q. Current overview of mushroom production in the world. In: Edible and Medicinal Mushroom: Technology and Applications,

SANTOS, H. The market mushroom mushroom in the country,2004. Available at:< http://g1.globo.com/economia/agronegocios/globo-rural/noticia/2018/07/cresce-mercado- mushroom-in-the-country.html>. Accessed on: 15 Jan. 2020 SIDRA, IBGE Automatic Retrieval System. Data from the 2006 Agricultural Census and the 2017 Agricultural Census.

SANTOS, J. R. et al. Mushroom cultivation in Bahia: an analysis of sustainable practices. Agropecuária Brasileira, Brasília, v. 20, n. 1, p. 45-58, 2020.

SILVA, Clóvis Gouveia da**.** Optimisation of the algaroba brandy production process and use of solid waste and food products. 2009. Dissertation (Master's in Process Engineering) - Federal University of Campina Grande, Campina Grande, 2009. SILVA, M.; RAMOS, A. C.; LIDON, F. J.; REBOREDO, F. H.; Gonçalves, E. M. Estratégias
pre- and post-harvest for Pleurotus ostreatus mushroom in a circular economy approach. Food, v. 13, p. 1464, 2024.

SILVA, Michelle Madureira e. Growing edible mushrooms using the Jun-Cao technique. Postgraduate course in Microbiology. Institute of Biological

Sciences, UFMG. 2011.

SINGH, Manjit; Kamal, Shwet; Sharma, V.P. Situation and trends in world mushroom production-III: world production of different mushroom species in the 21st century, 2020. SIQUEIRA, Carolina de Souza. Utilisation of agro-industrial waste in forest plantations. 2010. Final Course Work (Undergraduate Degree in Forestry Engineering) - Federal Rural University of Rio de Janeiro, Forestry Institute, Seropédica-RJ, 2010.

SIQUEIRA, F. G.; MACIEL, W. P.; Martos, E. T.; Duarte, G. C.; MILLER, R. N. G.; Silva, R.; Dias, E. S.. Cultivation of Pleurotus mushrooms in substrates obtained by short composting and steam pasteurisation. African Journal of Biotechnology, v. 11, p. 11630-11635, 2012.

SOUZA, A. B. et al. Plant populations and levels of fertilisation and liming for bean (Phaseolus vulgaris L.) in a low-fertility soil. Ciência e Agrotecnologia, Lavras, v. 26, n. 1, p. 87-98, 2002.

TEDESCO, M.J; Gianello, C; Bissani, C.A; Bohnen, H; Volkweiss, S.J. 1995. Soil analyses, UFSC, pg. 19-56.

URBEN, A.F. (Ed.). Mushroom production using modified Chinese technology: biotechnology and applications in agriculture and health. 3.ed. rev. and expanded. Brasília: Embrapa, 2017. WURMSTICH, I. S.; JUNIOR-SILVA, C. et al. Reuse of agro-industrial waste as a substrate for the production of edible macrofungi. Federal Institute of Education, Science and Technology of Mato Grosso, Journal of technological week, extraordinary edition, n. 05, 2021.

XIEMENES, Luciano Feijão; SILVA, Maria Sonia Lopes da; BRITO, Luiza Teixeira de Lima (technical editors). Technologies for coexistence with the Brazilian semi-arid region. Fortaleza, CE: Banco do Nordeste do Brasil, 2019. 1116 p. (BNB Science and Technology Series).

ZAKIL, A.F; ISA, M.R; Sue, M.S.M; Isha, R. Growth performance and mineral analysis of Pleurotus ostreatus (oyster mushroom) cultivated on spent mushroom medium mixed with rubber tree sawdust. Materials Today: Proceedings. vol. 57, p. 1329-1337, 2022., Z.; ZHANG, Y.; Liu, H.; WANG, J.; WANG; Deng, Z.;

LI, T.; He, Y.; YANG,.;

zhong, S. A water-soluble selenium-enriched polysaccharide produced by Pleurotus ostreatus: Purification, characterisation, antioxidant and antitumor activities in vitro. International Journal of Biological Macromolecules, v. 168, p. 356-370, 2021.

ZHENG, C; LI, J; Liu, H.; WANG, Y. Review of postharvest processing of edible wild-grown mushrooms. Food Research International, v. 173, p. 113223, 2023.

ZIED, D. C.; Pardo-Giménez, A. (Eds.), WILEY-BLACKWELL, West Sussex, p. 5-13, 2017.

ZIED, Diego Cunha; PARDO-GIMÉNEZ, Arturo; SÁNCHEZ, José Ernesto. Sustainability in mushroom production systems: a review of cultivation techniques and waste management. Agriculture, v. 11, n. 2, p. 98-110, 2021.

MIX
Papier aus verantwortungsvollen Quellen
Paper from responsible sources
FSC® C105338

Printed by Books on Demand GmbH, Norderstedt / Germany